How Many?
Counting to 5

A CRABTREE SEEDLINGS BOOK

Miranda Kelly

CRABTREE
PUBLISHING COMPANY
WWW.CRABTREEBOOKS.COM

Let's learn to count!

1   2   3

2

Counting tells us how many.

4   5

One dog

1

One cat

1

How many dogs
do you see?

Two hats

2

Two boots

2

How many boots
do you see?

Three turtles

3

Three frogs

3

How many turtles
do you see?

14

Four buckets

4

Four flip-flops

4

How many flip-flops
do you see?

Five eggs

5

Five candy canes

5

How many candy
canes do you see?

School-to-Home Support for Caregivers and Teachers

Crabtree Seedlings books help children grow by letting them practice reading. Here are a few guiding questions to help the reader with building his or her comprehension skills. Possible answers are included.

Before Reading

- What do I think this book is about? I think this book is about counting. It might show us different objects to count.

- What do I want to learn about this topic? I want to learn how to count.

During Reading

- I wonder why... I wonder why the turtles in the picture are holding each other.

- What have I learned so far? I have learned that I can use my fingers to count to five. I have five fingers.

After Reading

- What details did I learn about this topic? I learned the numbers one, two, three, four, and five.

- Write down unfamiliar words and ask questions to help understand their meaning. I see the word **flip-flops** on page 17. I can see sandals in the picture. Are flip-flops a type of sandal? Where do people wear flip-flops?

Library and Archives Canada Cataloging-in-Publication Data

Title: How many? : counting to 5 / Miranda Kelly.
Names: Kelly, Miranda, 1990- author.
Description: Series statement: Early learning concepts | "A Crabtree seedlings book". | Previously
  published in electronic format by Blue Door Education in 2020.
Identifiers: Canadiana 20200382624 | ISBN 9781427128430 (hardcover) | ISBN 9781427128515 (softcover)
Subjects: LCSH: Counting—Juvenile literature.
Classification: LCC QA113 .K45 2021 | DDC j513.2/11—dc23

Library of Congress Cataloging-in-Publication Data

Names: Kelly, Miranda, 1990- author.
Title: How many? : counting to 5 / Miranda Kelly.
Description: New York : Crabtree Publishing, 2021. | Series: Early learning concepts ; a Crabtree seedlings book
Identifiers: LCCN 2020049654 | ISBN 9781427128430 (hardcover) | ISBN 9781427128515 (paperback)
Subjects: LCSH: Counting--Juvenile literature.
Classification: LCC QA113 .K445 2021 | DDC 513.2/11--dc23
LC record available at https://lccn.loc.gov/2020049654

Crabtree Publishing Company
www.crabtreebooks.com   1-800-387-7650
e-book ISBN 978-1-947632-89-9
Print book version produced jointly with Crabtree Publishing Company NY, USA

Written by Miranda Kelly
Production coordinator and Prepress technician: Amy Salter
Print coordinator: Katherine Berti

Printed in Canada/022022/CPC20220214

Photo Credits: istock.com, shutterstock.com, Cover: By LukaKikina;
Pg1; shutterstock.com/ LukaKikina. Pg2/3; shutterstock.om/LeventKonuk. Pg4/5; istock.com/GlobalP, istock.com/MirasWonderland. Pg6/7; istock.com/Kerkez. Pg8/9; mawielobob, MirasWonderland. Pg10/11; istock.com/Hakase_. Pg12/13; istock.com/amwu, istock.com/ithinksky. Pg14/15; istock.com/cturtletrax. Pg16/17; istock.com/filipfoto/filipfoto. Pg18/19; istock.com/Nadezhda1906. Pg20/21; istock.com/PhotoMelon, istock.com/robynmac. Pg22/23 istock.com/LeszekCzerwonka. Pg24; shutterstock.com/ Sichon

Published in Canada
Crabtree Publishing
616 Welland Ave.
St. Catharines, ON
L2M 5V6

Published in the United States
Crabtree Publishing
347 Fifth Ave
Suite 1402-145
New York, NY 10016

Published in the United Kingdom
Crabtree Publishing
Maritime House
Basin Road North, Hove
BN41 1WR

Published in Australia
Crabtree Publishing
Unit 3 – 5
Currumbin Court
Capalaba QLD 4157